ElementOP

Max Freedom Pollard

ISBN 978-1383835670

ACKNOWLEDGMENTS

This book is dedicated to my mother Lily Moussa.

Special thanks to Frans de Waal (Yerkes National Primate Research Center, Atlanta) and Katarzyna Nowak (Mammal Research Institute, Białowieża).

I would like to acknowledge the Friends of the Library Committee at the University of Adelaide.

CONTENTS

PROPOSED MODELS

This book posits these new theories:

That imprinting exists in all mammals and advanced species.

That the evolutionary advantage imprinting provides is an advantage to evolution itself, facilitating evolution in relation to object type, as opposed to form.

That ethological imprinting processes are identifiable in humans. Three are proposed.

DEFINITIONS

Alpha
Used strictly to denote a high-ranking member of a primate troop, unrelated to the term as used in pop psychology and common culture.

Circuit
Convenience term for the subset of a nervous system modulated by an imprint.

Engram
Convenience term for a unit of cognitive information, persistently and retrievably recorded in a nervous system.

Non-Reproductive Orientations
Sexual orientations not oriented toward reproductive pairings, without implying any inability to reproduce.

Troop
A social group of primates.

UNDERSTANDING IMPRINTS

Ethological imprinting is a straightforward phenomenon to identify and observe in animals. In fact, ethological imprints were initially discovered based upon simple observation of animal behavior.

Three criteria define ethological imprinting:

Sensitive period
A genetically timed window determines the period of imprint vulnerability, during which an imprint can occur.

Aspect-identifying stimulus
A specific type of stimulus from the environment, representing object aspect. Stimulus appearance during the sensitive period identifies the object to be imprinted.

Persistent downstream effects
The imprint of an object persists, and is utilized by the nervous system for object-oriented behavior and perception.

Imprinting was first discovered in birds because their imprints are easiest to observe.

In geese, the window for maternal imprinting is brief and genetically timed, occurring about 13-16 hours after hatching. Any moving object near a gosling during that period can meet the stimulus criteria and become the object of imprinting. From that moment onward, the gosling recognizes whichever moving object it encountered during that three-hour window as its "mother" and will follow it through the early developmental period.

Early ethologists discovered that any object[1] introduced 13-16 hours after a gosling's birth could become the target of imprinting: the gosling would imprint on it and recognize it as its mother. Geese have been observed imprinting on other species, remote-controlled cars, boots, and many other objects simply because those objects were placed near the gosling during this sensitive period. Even if the real goose mother is introduced immediately afterward, a gosling that has imprinted on a non-goose object will keep following that object rather than her.

Imprinting in geese establishes an imprint of the mother, and this imprinted engram irreversibly determines the perception of 'mother', permanently altering behavioral perception. The form of the mother, or whichever object

[1] Any object larger than the gosling, but small enough to be wholly observable by the gosling, that also moves.

is imprinted as the mother, is stored as an engram that persists into adulthood. This engram appears to be referenced by a goose nervous system in various ways, depending on developmental stage.

After development, as adults, male geese actually become sexually oriented toward the imprinted object, the 'mother image.' The mother imprint is referenced by the nervous system as the form of a reproductive female in their particular place and time. Males direct their mating efforts toward objects that fit the imprinted "mother image," no matter how atypical.

Imprinting isn't strictly age ascending (young→adult), it can be top-down (adult→newborn) or lateral (same-stage↔same-stage). For example, the imprinting window for offspring recognition[2] in ewes lasts around two hours immediately after birth. The imprint stimulus is the olfactory signature of the ewe's postpartum placenta. The result is a persistent maternal engram of the lamb's individual features, after which the ewe accepts only that lamb, and rejects others.

Farmers use this imprint to graft orphan lambs onto different ewes, in a process called slime grafting - rubbing the orphan lamb with the foster ewe's placental fluids immediately after the ewe gives birth. Farmers can graft sheep onto goats and vice versa, using slime grafting, as they share the same recognition imprint.

---

[2] The mother recognizing the child, not the child recognizing the mother.

In salmon, the first imprint results from olfactory cues. A brief sensitive phase opens in juvenile salmon as they leave their natal spawning stream. The stimulus is the river's distinctive blend of odorants. This establishes a stable olfactory engram that enables adult salmon to navigate complex river systems back to their natal spawning grounds, where they lay their eggs.

In sea turtles, imprinting is geomagnetic. The imprinting window occurs just after hatching, when hatchlings first reach the ocean. The stimulus is the ocean they encounter right after entering the water. The imprint is of the local geomagnetic signature, resulting in a persistent engram that guides later orientation and movement, enabling adult females to return to the same coastline decades later to nest.

Imprinting can shape more than caregiver recognition and sexual behavior. Imprinting is not confined to development, it is development.

The cuckoo and its hosts provide a notable example of imprinting. In certain birds, parents imprint on the first chick, which serves as the recognition template for later hatchlings. Every hatchling thereafter is judged against the engram of the first, and later hatchlings whose appearance or behavior fail to match it are rejected.

The imprinting aspect of "first-hatched," used to identify offspring, stems from an evolutionary near certainty that the first hatchling in a nest would belong to the breeding pair that built the nest. This parallels the evolutionary assumption that the lamb belonging to a ewe would be the one coated in the ewe's placental fluids, or that the best representation of a safe spawning site in a salmon's particular space and time, is the one it was successfully spawned in.

So many birds imprint on the first-hatched chick for offspring identification, that brood parasites such as cuckoos have evolved to exploit this biological reality.

Cuckoos are not subtle deceivers of their hosts. A reed warbler may weigh about as much as a teaspoon of sugar, yet the cuckoo chick it raises can dominate the nest even while still a fledgling, growing to nine times the weight of the adult reed warbler that feeds it.

This almost comic deception is striking. As Bruce Lyon, professor of ecology at UCSC, remarks, "When you see a little songbird struggling to feed an enormous cowbird[3] chick, you have to wonder why it cannot recognize the parasitic chick when it is so obvious to us."

Why did birds evolve to imprint on their first hatchling? Why didn't they evolve to raise every hatchling, or simply evolve static recognition of their own kind?

Imprinting exists to prevent lineage-terminating errors, such as rejecting a hungry chick because it looks different from the historical average. Without imprinting, any mutation that results in a perceptible change in a single bird would be hit with an evolutionary difference-penalty

[3] Cowbirds practice similar brood parasitism.

in any process involving another member of the species, such as in parental recognition and mate selection, with the difference-penalty increasing with degree of difference.

The reason a bird is better off[4] imprinting upon its young, is that a new hatchling might have evolved in a perceptible way, and rejecting all perceptually different hatchlings would permanently end evolution in that lineage.

A bird whose nervous system has evolved certain behaviors in relation to a fixed 'hatchling form' cannot use those behaviors in relation to any chick that differs noticeably from that hatchling form, unless a matching mutation allows recognition of the separately mutated bird.

Simultaneously occurring mutations to recognize other mutations generally isn't feasible. A species without the imprinting process would struggle to evolve in a perceptual way, as any perceptible mutation would result in a difference-penalty.

If an ordinarily red chick was to be born blue in a world without imprints, its parents would need a separate mutation to accept the blue bird. Future mates would need another separate mutation to recognize the blue bird as a suitable mate, and experience sexual desire towards it. This deviation from the norm would end the

---

[4] From an evolutionary perspective, not generally.

blue bird's evolutionary lineage, even if being born blue better suited the environment, and otherwise conferred advantage.

That would turn every perceptible mutation into a detrimental mutation. The evolutionary advantage of imprinting isn't towards an individual animal, but to the evolutionary lineage itself. The evolutionary benefit is that it facilitates evolution.

By allowing an animal's nervous system to evolve in relation to core environmental objects that remain the same in type - mother, mate - but whose forms will necessarily change, imprinting creates a kind of reproductive algebra, allowing environmental objects to be genetically anticipated even if their future form is unknown. Imprinting allows species to accept beneficial mutations in a single generation, and evolve in a way accepting of future evolution.

Any animal species with an imprinting process can evolve much faster than one without, by removing difference-penalties for mutations, penalties that would otherwise come from the mother/father/pack/species. Imprinting is an adaptation that allows further adaptations to exist.

Without imprinting, a novel trait that changes how an animal looks, sounds, or smells starts life at a disadvantage purely because it's rare.

Imprinting is fundamental, and resultant engrams are prioritized by the nervous system over conceptual logic, as the engram "fills a spot" which nervous systems have evolved in relation to, and thus requires.

These spots cannot be filled by conceptual intelligence and conceptual recognition. Conceptual intelligence sufficient for human-style 'conscious' perception of objects, and choice of action in relation to those objects, varies across species, but even in species which possess this ability, action in relation to key environmental objects is required before those mental facilities develop, and the parts that make up conceptual intelligence developed as a sub-processes of the nervous system, entirely within the context of, and modulated by, older neural structures and their intended purposes.

A bird targeted by a brood parasite can't "think" its way out of the initial deception, because the effect of an imprint on behavioral perception is fixed. Imprints can be understood as a kind of animal firmware - stored on hardware, fundamental to software, and only editable during a specific time phase with specific input.

During a short window, a reed warbler's brain latches onto an engram for what counts as "its hatchling." Once that template is imprinted, all cognition thereafter references the engram of the hatchling as the hatchling itself. To an imprinted reed warbler this perception is as real, perhaps more real, than its normal identification of physical objects.

If mate choice is genetically encoded towards a singular form, all new morphs suffers a mating penalty until a matching "preference-for-the-beneficial-trait" allele arises and spreads in parallel. That's a coordination bottleneck. The trait must wait for the preference to co-evolve. Most new morphs would die out in this waiting room, even if mutations offered survival or display benefits, because rarity alone depresses their mating, socializing and early-life opportunities. This explains the existence of imprinting processes across unrelated taxa.

The need for social recognition of benefit, for benefit realization in a social group context, is clear. If someone was to develop and market a *better* product, but nobody recognized that product as better, the product does not realize any benefit at all, and will not persist as a product.

The same need for social recognition of benefit is true for a beneficial mutation in an animal. If the benefit results in rejections by others of the same species, the benefit is never realized via increased reproduction, and so the mutation fails to persist. Imprinting facilitates benefit realization through the removal of difference-penalties.

Genes and their physical manifestations don't store an object list, they store parameters to identify environmental objects in an animal's particular space and time.

An environmental object of a certain type is identified by aspect, and when the stimulus associated with the object is received during the sensitive period, identification of the object results in recording of its current form as an engram. This form is utilized by the nervous system for whatever purposes the animal might have evolved in relation to the object.

Thanks to imprinting, genes can evolve in relation to anticipated environmental variables, with the environment supplying the actual values.

In this manner, an animal can realize mutational benefit without others of its species having synchronous mutations to allow recognition.

Engram Architecture

Nervous system architecture is complex, and physical identification of engram form and utilization is beyond the scope of this book. As "On the Origin of Species" described evolution conceptually a century before genes were first proposed, this book necessarily uses many convenience terms and focuses on observable input and output. All attempts in this book to identify the specific neurological changes that constitutes an imprinting process are just that, attempts.

Reproductive Algebra

Imprinting moves reproduction past fixed values, allowing reproductive algebra, an enormous advantage over more static evolutionary lineages. In a fixed-preference model $P = v$ (v being a genetically encoded constant): changing v requires a new mutation. With imprinting, no specific value is stored; what's stored is the rule for setting it: during the sensitive window τ, set $P \leftarrow f(X\tau)$, where X is the observed attribute set and f is the allele-encoded mapping. This allows cognitive and reproductive systems to adapt in relation to object type, not form, allowing adaptation to what will exist, not merely what does.

Evolutionary Benefits

With imprinting, populations don't need to carry and synchronize dozens of preference alleles for every possible morph. The same developmental script flexes to whatever the cohort contains. Representation of an evolutionary object during the imprint window, not historical frequency, establishes the form of evolutionary object which survival and reproduction systems are oriented to.

Imagine a mutation results in a white pup being born into a pack of black wolves on snowy terrain. Without imprinting, juveniles are genetically oriented to the established form of their species, black wolves of a familiar size and shape. The white morph pays a reproductive penalty in relation to the degree of departure from existing morphs, despite its improved adaptation to a snow-white environment. Unless a separate "prefer white" bias evolves concurrently, acceptance of the trait won't scale in proportion to its benefit.

Pups that repeatedly see, hear, and smell the white adult during the imprint window can store white as part of their mate recognition parameters. If the white morph survives or hunts better in snow, its prominence in the pack makes it more likely for pups to imprint upon. In a single generation, the beneficial mutation translates into higher mating probability.

If the white trait confers even a modest advantage, its frequency in a population can rise without the usual handicap of difference-penalty, and the next cohort encounters more white adults. Trait benefit and preference climb together in a tight spiral.

IMPRINTING IN INDUSTRY

In first century Rome, the agricultural writer Lucius Moderatus Columella advised that anyone who wished to raise ducks should collect eggs from the marshes and place them under farmyard hens. Raised in this way, he said, the ducklings would shed their wild nature.

Columella intuitively grasped the idea of filial imprinting. Today, imprinting informs practices in wildlife management, aquaculture, aviation, farming, and other fields that work with animal behavior.

Wildlife managers use filial imprinting, guiding chicks with puppets and carefully chosen cues to affix their natural instincts upon the desired object. Field programs apply the same principle across entire landscapes. The Italian pilot Angelo d'Arrigo led hand reared raptors along migration routes after imprinting cranes on his glider during early handling, and the birds then followed his aircraft for thousands of kilometers.

Falconry is a living heritage in Saudi Arabia and across the Middle East, founded on imprinting to serve human aims. Many falconers hand rear hawks, producing tame birds that bond closely with their keepers, fly where directed, and return. Humans have hunted and farmed this way for thousands of years, with Chinese farmers imprinting ducklings onto sticks so an entire flock can be directed across rice paddies.

In mammals, imprinting is rarely as straightforward as the "follow the first moving thing" style known in birds. Instead, it is often olfactory and multisensory, centered on bonding between parent and offspring during a brief sensitive period immediately after birth.

Zoos around the world use imprinting to support breeding programs.

At the Wolong panda reserve, staff limit human contact and even wear panda suits around cubs so the young do not imprint on people. Human focused imprinting can suppress normal courtship between pandas.

Occasionally an unintended imprint disrupts breeding plans. Chi Chi, the female giant panda at the London Zoo, seemed to imprint on humans. When she was sent to the Moscow Zoo to mate with the male An An, she rejected him but offered a full sexual display to a zookeeper.

Ewes, like many hoofed mammals, have a very brief and critical window for imprinting. In the first four hours after birth, a ewe forms a highly selective bond with her lamb, guided mainly by smell. She learns the newborn's unique scent almost at once and will accept only that lamb, rejecting others of the same age and appearance. This ensures that her care goes to her own offspring and prevents misdirected nursing.

Farmers use this imprint to manage sheep and other livestock. Goat herders graft orphan kids onto foster does, and sometimes onto postpartum animals of related species such as sheep, by rubbing the orphans with the foster's placental fluids so they are accepted. Because many hoofed livestock recognize their young by the scent of a postpartum mother's placental fluids, orphans can be reallocated to foster mothers, even across species.

Experiments have even hinted at prenatal imprinting to predators in rodents. When pregnant rats encounter predator scents, their pups are born with a permanent aversion to those olfactory cues. The young arrive with an updated avoidance response that fits the dangers present in their particular place and time, although it's debatable as to whether that constitutes an ethological imprint.

Fisheries and aquaculture managers deliberately engineer natal-homing imprinting to control where fish return. Juveniles are reared in hatchery water with a distinct chemical "signature," then held in acclimation raceways or imprinting ponds at the intended release point for days so those olfactory cues are imprinted. The result is higher return rates to the correct river or facility, enabling targeted harvest and broodstock collection while reducing genetic and ecological risks from straying into wild spawning grounds.

Livestock guardian dogs are bonded to sheep or goats during a narrow puppy window between four and sixteen weeks. This produces adults that prefer, follow, and defend livestock rather than other dogs. Many ranch extension guides describe this as an imprinting practice.

Imprinting is documented in dogs, sheep, ducks, geese, chickens, finches, goats, cranes, voles, fishes, turtles, dolphins, and even some invertebrates.

Imprinting is not merely of academic interest, manipulation of imprint processes are already vital to countless industries. This utilization will only increase and improve alongside further discovery.

IMPRINTS AS SYNAPOMORPHY

A synapomorphy is a shared trait that originated in a group's common ancestor, that is inherited by all of its descendants. Such traits identify and unite the group.

In mammals, synapomorphies include not only anatomical traits such as the neocortex and hair, but also process synapomorphies: genetically programmed processes that enable core mammalian functions. Internal fertilization, lactation and weaning, passive immune transfer, sleep, and puberty and sexual maturation are examples shared by all mammals

Lactation is a process synapomorphy: all mammals produce milk via mammary glands to nourish offspring at the appropriate time. Endothermy is another, enabling mammals to generate internal heat to regulate body temperature and thrive across diverse environments. Sleep is a conserved, cyclical state of reduced responsiveness that structures the daily rhythm of every mammal.

Imprinting is part of the shared evolutionary heritage of the class Mammalia and should be treated as a mammalian synapomorphy, not as a niche or ambiguous descriptor. The primacy of imprinting can be inferred from common descent and divergence points.

If both sides of the key splits show bona fide imprinting (narrow window, input-specific engram, durable template), then the minimum origin of imprinting is at least as old as that split. Mammals, as descendants, inherit the program. By mapping species with empirically confirmed imprinting processes onto a phylogenetic tree, we can reconstruct the trait's evolutionary history, including its origins and any convergent evolutions.

Imprinting is less an evolutionary beneficial mutation, and more of an evolutionary requirement for continued evolution in higher order animal lineages. It's the indispensable trait from an evolutionary perspective.

The claim that imprinting processes are an inherent trait across all mammals is a bold one, but is evidenced via an examination of the estimated dates of species divergence between animals in whom an imprinting process has been clearly established.

This allows the identification of fixed points in time where it can be assumed that all predecessors share the same process or similar, insofar as both the first and last members of divergent branches share the same trait despite otherwise dissimilar traits, with a distribution as diverse as other mammalian traits like a neocortex or immune transfer.

In order to compare multiple emergence of the imprinting process in mammals, we'll create some divergence anchors for the purposes of comparison.

ANCHORS

We seek to demonstrate the minimum origins of imprinting processes in various species by using anchor points from separate lineages to triangulate when the trait first appeared.

If imprinting is present in different groups that split early in mammalian history, the most parsimonious explanation is that the trait was present in their last common ancestor. Each additional convergence greater implies shared origin. The casual reader should skip this chapter, which is necessarily verbose and repetitive.

Anchor 1: Fishes vs. Tetrapods (Actinopterygii ↔ Sarcopterygii → Tetrapoda; Devonian ≥ 420-450 Ma)

Ray finned fishes (Actinopterygii)

➤ Pacific salmon (Oncorhynchus spp.): narrow juvenile windows, often opening at alevin or parr stages, write natal river odors into a permanent engram, adults return with meter scale fidelity. The imprint write retunes the olfactory pathway, including receptor expression and sensitivity. Hatcheries use these windows with controlled odor exposure.

➤ Chinook, Coho, Sockeye, Chum: parallel olfactory imprinting programs; juvenile exposure to specific odorants such as phenethyl alcohol biases adult homing years later.

➤ Lake sturgeon (Acipenser fulvescens): early life olfactory imprinting during the free embryo or larval stage; used operationally in reintroduction programs.

➤ Coral reef fishes such as clownfish and damselfish: pelagic larvae acquire olfactory engrams of natal reef or anemone odors within a short window before settlement; later choose home reefs accordingly.

➤ Geomagnetism in fishes: salmonids show evidence of geomagnetic imprinting for ocean scale navigation, with olfaction taking over near river mouths; two privileged channels coordinated across ranges.

Tetrapods (descendants of Sarcopterygii)

➤ Amphibians (see Anchor 2): larval windows for olfactory engrams of natal ponds that guide post metamorphic orientation.

➤ Birds (see Anchor 3): tight filial and sexual imprinting windows; acoustically gated templates in colony breeders.

➤ Reptiles (see Anchor 3): sea turtles encode geomagnetic natal signatures and later return.

➤ Mammals (see Anchors 3 to 5): perinatal olfactory, and sometimes acoustic, windows write selective social templates.

Anchor Summary:

Ray finned fishes pass through brief juvenile imprinting periods during which natal odors are encoded as durable homing engrams. In salmon, this period even remodels the peripheral sensory system, including up-regulation of odorant receptor genes and increased sensitivity of guanylyl cyclase. Sturgeon, which are distantly related to salmonids, show similar larval olfactory imprinting, confirming that this process is not unique to salmon.

Tetrapod descendants likewise show imprinting across multiple senses, including olfactory, acoustic, and geomagnetic cues. Because both Actinopterygii and tetrapods share three key features a short developmental window, a privileged sensory pathway, and a durable memory engram. The mechanism likely arose before the split between Actinopterygii and Sarcopterygii in the Devonian, roughly 420 to 450 million years ago, rather than being a late innovation in terrestrial vertebrates.

Anchor 2: Amphibians vs. Amniotes (Amphibia ↔ Amniota; late Devonian/Carboniferous ≥ 350-320 Ma)

Amphibians

➤ Salamanders and newts (Ambystoma, Triturus): brief larval sensitive periods for learning natal pond odor;

adults later show natal site fidelity guided by that stored engram.

➤ Anurans, meaning frogs and toads: tadpoles exposed to specific chemical signatures during narrow larval windows later orient to those cues for refuge choice, microhabitat selection, or breeding site return.

Amniotes, meaning sauropsids and synapsids

➤ Birds: classic filial imprinting in ducks and geese within hours of hatching; sexual imprinting in species such as zebra finches and quail during tight juvenile phases that shift adult mate choice; many passerines have narrow song learning windows, and acoustically gated engrams are a hallmark in colony breeders.

➤ Reptiles, marine: sea turtles including loggerhead, green, and leatherback acquire a natal geomagnetic address early in life and use a geomagnetic map to return years later; hatchlings also show brief olfactory sensitivity to beach odors. As in salmon, magnetism supplies a broad scale map, with olfaction taking over near the target.

➤ Mammals: perinatal olfactory windows in sheep and goats enable rapid, selective bonding between mother and young; blocking beta adrenergic and noradrenergic input to the main olfactory bulb during the window prevents bonding. In neonatal rats, a short sensitive period is driven by a hyperactive locus coeruleus, norepinephrine floods the bulb, engages plasticity switches such as pCREB, and records a durable maternal odor template that guides approach. Parallel early life olfactory and acoustic plasticity appears across rodents, bats, pinnipeds, and cetaceans.

Anchor Summary:

Both amphibians and amniotes exhibit true imprinting. There's a brief sensitive window that writes a template through a privileged channel, yielding a durable, behavior-shaping memory. Because the full imprinting triad is present on both sides of the Amphibia-Amniota split, the origin must predate the amniote radiation, forcing a minimum age in the late Devonian/Carboniferous (≥ 350-320 Ma), not a late innovation of birds or mammals.

Anchor 3: Sauropsids vs. Synapsids (Amniota split; late Carboniferous ≥ 315-320 Ma)

Sauropsids

➤ Birds, filial: ducklings and goslings follow the first appropriate moving object during a tight post hatch window, privileging visual and acoustic channels.

➤ Birds, sexual: zebra finches cross fostered to other species later prefer foster type mates; a narrow juvenile window sets mate choice templates.

➤ Birds, acoustic identity: penguin parents and chicks in dense colonies imprint each other's calls; these vocal templates remain robust even in intense masking noise.

➤ Birds, navigation: pigeons and other migrants calibrate geomagnetic and olfactory maps during early life; orientation learning is time locked to a sensitive developmental phase.

➤ Reptiles, sea turtles: hatchlings encode natal geomagnetic signatures in an early window; juveniles later combine that imprint with oceanic cues to target the correct coastline.

➢ Shared feature: across birds and turtles, different channels, visual, acoustic, olfactory, and geomagnetic, all instantiate the same triad: brief sensitive window → privileged sensory channel → durable navigational or social template.

Synapsids

➢ Ungulates such as sheep and goats: within one to four hours postpartum, an olfactory gate flips ewes from general newborn attraction to selective recognition of their lamb; blocking noradrenergic and beta adrenergic signaling in the main olfactory bulb during this window abolishes selectivity. Mechanism: the postpartum gate is explicitly noradrenergic, and beta receptor blockade prevents formation of the ewe and lamb bond.

➢ Rodents such as rats and mice: a neonatal locus coeruleus norepinephrine surge opens a short olfactory window; pairing a novel odor with gentle tactile stimulation writes a durable approach template via pCREB tagged plasticity. Mechanism: the norepinephrine surge primes plasticity switches in the olfactory bulb, and coupling with sensory input creates an imprinted template within minutes.

➢ Pinnipeds such as fur seals and sea lions: mothers and pups rapidly establish mutual vocal identities within hours postpartum; olfactory checks provide secondary verification at close range.

➢ Cetaceans such as dolphins: mothers and calves rely on individualized signature whistles; mothers exaggerate acoustic features in calf directed motherese, a pedagogical adaptation that supports imprinting.

➢ Bats: pups learn maternal acoustic signatures and roost odor profiles within a narrow postnatal window; mothers quickly recognize isolation calls.

Anchor Summary:

Both sauropsids and mammals operate robust imprinting programs. Because the full imprinting triad is present in both sauropsids and synapsids, the capacity must be at least as old as the amniote split (≥ 315-320 Ma). Convergent reinvention across both branches would require independent evolution of the same gated circuit architecture, which is far less parsimonious than deep homology.

Anchor 4: Metatherians vs. Eutherians (Theria split; Middle-Late Jurassic ≥ 160-170 Ma)

Marsupials, meaning Metatheria

➢ Tammar wallaby and kangaroos: neonates are born extremely altricial and use olfaction to locate the teat and pouch; pouch and teat odors are privileged, and early attachment rapidly becomes selective. Mechanism: extreme altriciality forces olfactory privilege, with semiochemicals in maternal secretions providing the scaffold, and the attachment window closes quickly as selective bonds form.

➢ Opossums: day old young show directed chemotaxis to maternal odors and nipple cues; disrupting cues during the immediate window derails attachment. Mechanism: the same marsupial pattern of olfactory privilege applies, with maternal secretions driving a short attachment window that closes quickly.

➢ Koalas: joeys pass through a brief period in which

exposure to maternal pap and its odor and taste profile scaffolds a long term diet template for eucalyptus selection. Mechanism: maternal secretions and semiochemicals provide the template during a short marsupial window that rapidly becomes selective.

Placentals, meaning Eutheria

➤ Artiodactyls such as sheep, goats, deer, and cattle: postpartum olfactory windows yield selective bonding between mother and young; attraction to amniotic fluid flips from general to selective within hours, showing a prewired gate rather than slow learning. Mechanism: postpartum noradrenergic gating in olfactory circuits, including beta adrenergic dependence at the bulb, creates a time locked opportunity for template writing, and amniotic fluid and colostrum carry critical cues.

➤ Rodents such as rats, mice, and voles: neonatal olfactory imprinting on maternal and pup odors; early exposures bias later social and mating preferences. Mechanism: a neonatal locus coeruleus norepinephrine surge primes plasticity switches in the olfactory bulb, and pairing sensory input with that state writes a durable template within minutes, including pCREB tagged plasticity in classic preparations.

➤ Carnivora such as pinnipeds, canids, and mustelids: rapid vocal and olfactory recognition between mother and young supports reunions in crowded rookeries or enclosed dens. Mechanism: early sensitive periods in olfactory and acoustic circuits operate within the general eutherian framework of noradrenergic gating for rapid bonding.

➤ Chiroptera, meaning bats: early acoustic and olfactory

learning stabilizes mother and pup reunion in dark, dense roosts. Mechanism: early sensitive periods in both channels allow rapid template formation consistent with eutherian noradrenergic gating.

➤ Cetacea such as dolphins and some baleen whales: individualized acoustic identity, for example signature whistles, is established in early calfhood. Mechanism: early acoustic plasticity provides the template for recognition and reunion in a sensory domain that is primary for marine mammals.

➤ Primates such as humans, macaques, and marmosets: newborns show immediate biases for maternal voice and odor; brief windows for face and voice tuning and speech perceptual narrowing set long lived social and perceptual templates. Mechanism: early sensitive periods in auditory, visual, and olfactory systems support template writing, consistent with eutherian noradrenergic gating for rapid selective bonding..

Anchor Summary:

Both marsupials and placentals run bona fide imprinting programs. Because this full triad spans both metatherian and eutherian branches, the capacity predates the therian split (≥ 160-170 Ma), arguing for deep homology rather than repeated convergence.

Anchor 5 — Across placental Orders (multiple independent radiations inside Eutheria)

Artiodactyla/Carnivora

➤ Artiodactyla, such as sheep, goats, and cattle: postpartum olfactory windows yield selective bonds between mother and young. Neonatal cross fostering between sheep and goats can later shift adult mate preference toward the foster species, a direct mammalian example of a durable sexual imprint. Mechanism: a brief postpartum gate in olfactory circuits, strongly noradrenergic and supported by oxytocin pathways, writes the template when amniotic fluid and early secretions carry the cues.

➢ Carnivora, such as fur seals, sea lions, and canids: in pinnipeds, vocal identity learned within hours supports reunions in noisy rookeries, with olfactory confirmation at close range. Canids imprint on den and natal range odours, biasing later dispersal and site fidelity. Mechanism: time locked windows privilege acoustic channels in pinnipeds and olfactory channels in canids, with noradrenergic state enabling rapid circuit level plasticity.

Rodentia/Chiroptera

➢ Rodentia: neonates form olfactory templates for maternal odor, and dams show a short plasticity window for rapid learning of pup calls. Early diet or odor exposure can bias later mate choice in some species. In prairie voles, oxytocin and vasopressin receptor topography, including V1aR, gates partner preference, so altering the map shifts bonding. Mechanism: a neonatal locus coeruleus norepinephrine surge primes pCREB tagged plasticity in olfactory and auditory pathways, while peptide receptor maps set the rule and exposure supplies the value.

➢ Maternal call plasticity in mice: oxytocin pathway neurons enable a time gated change that supports recognition of pup calls, tying imprint like learning to identified circuits and synapses. Mechanism: oxytocin driven modulation opens a short window that stabilises new responses in auditory cortex and related nuclei.

➤ Chiroptera, meaning bats: mother and pup call signatures and roost odor profiles are learned early, and juveniles acquire colony dialects during a narrow window. Mechanism: coordinated acoustic and olfactory sensitive periods create durable templates for reunion in dark, dense roosts, with noradrenergic state facilitating consolidation.

Cetacea/Primates

➤ Cetacea, such as dolphins and some baleen whales: signature whistle templates form in early calfhood, and mothers use exaggerated calf directed motherese that scaffolds the acoustic window for durable social identity. Mechanism: an early auditory sensitive period privileges fine scale spectral and temporal cues, allowing rapid stabilisation of a social identity template.

➤ Primates, such as humans, macaques, and marmosets: infants immediately privilege maternal voice and odor, and brief early windows for face and voice tuning and phoneme narrowing canalize long lived recognition and communication preferences. Mechanism: short, developmentally timed gates in auditory, visual, and olfactory systems, influenced by oxytocin and noradrenergic tone, write long lasting social and communicative templates.

Anchor Summary:

Imprinting is not a quirk of any one lifestyle. It reappears across independent placental radiations. Artiodactyla, Carnivora, Rodentia, Chiroptera, Cetacea, Primates, always showing the same architecture. This pervasive, rule-driven pattern marks imprinting as a conserved process trait within Eutheria, tuned by niche-specific sensory channels rather than a fixed list of species-specific preferences.

Anchor 6 — Deep Outgroup Bracket (Protostomes ↔ Deuterostomes; ≥ 600 Ma)

Protostomes

➢ Nematode (C. elegans): a brief larval olfactory window installs lifelong, odor specific preferences; the AWC to AIY pathway and receptor expression shifts are mapped; repeated early exposure can produce multigenerational persistence. Mechanism: named sensory to inter-neuron links, especially AWC to AIY, gate a short write window that stabilizes the template.

➢ Ants and wasps: callow adults pass through a short chemical window and learn the colony cuticular hydrocarbon blend that sets a lifelong nestmate template. Mechanism: early exposure tunes antennal lobe and higher olfactory circuits to the colony blend, which then anchors acceptance and aggression.

➢ Butterflies (Bicyclus anynana): newly eclosed females have a visual window for male wing pattern templates; early experience locks in adult mate preference. Mechanism: an early visual gate calibrates pattern sensitive circuits so later choice follows the stored template.

➢ Honeybees: young workers show an early olfactory learning bias that shapes later foraging and social responses to hive odors. Mechanism: reinforcement through octopamine and mushroom body plasticity consolidates odor templates during a short early window.

➢ Cephalopods such as cuttlefish: embryos exposed in ovo to prey or predator odors later show durable prey choice shifts and lateralized responses, meaning the write occurs before hatching. Mechanism: embryonic olfactory pathways encode prey or predator signatures that bias later approach or avoidance.

Deuterostomes

➢ Vertebrates, including fishes, amphibians, sauropsids, and mammals: multiple channel specific programs operate across taxa; see Anchors 1 to 5. Mechanism: short, time locked gates often shaped by noradrenergic like arousal systems and nonapeptides, with CREB family plasticity markers stabilizing the write.

➢ Echinoderms and tunicates: larval settlement often requires brief exposure to specific chemical signatures from conspecifics or habitat biofilms, yielding long lasting habitat choice templates. Mechanism: chemosensory detection of habitat cues triggers a one way switch into the settled state, fixing the habitat template..

Anchor Summary:

Because both protostomes and deuterostomes show narrow-window, channel-privileged, durable templates, the minimum origin of imprinting extends to at least the protostome-deuterostome split, and most parsimoniously to the bilaterian ancestor (570-650 Ma). The recurrence of the same gated architecture and its reliance on conserved switches (NE-like arousal systems, nonapeptides, CREB-mediated plasticity), argues for deep homology rather than repeated, independent reinventions.

Anchor Implications

At each deep branch - fishes vs. tetrapods; amphibians vs. amniotes; sauropsids vs. mammals; marsupials vs. placentals; protostomes vs. deuterostomes - both sides carry "true" imprinting. Each agreement moves the minimum origin backward to at least that divergence, making a late, lineage-specific invention increasingly implausible and a conserved, inherited program enormously the most likely interpretation.

Common ancestry

One may always claim "convergence," but the burden grows with each bracket. Our anchors stack like this:

Fishes ↔ Tetrapods both imprint → the rule is at least pre-Devonian (≥ 420-450 Ma).

Amphibians ↔ Amniotes both imprint → the rule predates amniotes (≥ 320-350 Ma).

Sauropsids ↔ Mammals both imprint → the rule sits at the amniote ancestor (315-320 Ma).

Marsupials ↔ Placentals both imprint → the rule is pre-therian (160-170 Ma).

Protostomes ↔ Deuterostomes both imprint → the rule stretches back toward the bilaterian ancestor (commonly bracketed 570-650 Ma).

Every "yes" on both sides ratchets the minimum origin back. By the time you include protostomes (nematodes, insects, cephalopods) and deuterostomes (all the vertebrate cases), convergence would require multiple independent evolutions of the same developmental logic - narrow window, neuromodulator/hormone gate, channel privilege, CREB-like write signals across lines that split in the Precambrian. That's a very tall order. The simpler story is a single ancestral program, with normal variation.

Specific Shared Machinery

Across clades you see the same class of switches turning the window on and off (noradrenergic/β-adrenergic gating in neonatal mammals; nonapeptide families like oxytocin/vasopressin and their vertebrate relatives; CREB-mediated transcription as the "write" signal). The channels differ by ecology and sensory input - olfaction on land, acoustics in colonies or underwater, geomagnetism offshore - but the meta-architecture in the nervous system is fundamentally the same: open the gate briefly, weight one channel heavily, write a template that is hard to reverse.

The similarity or lack thereof of sensory pathways between species is irrelevant. Just as every other genetic process such as lactation or sleep has necessarily adapted to the different physicality and sensory input of various species, so to, the imprinting process is shared among mammals despite varied physicality. Lactation is no less a shared trait among mammals, by virtue of breasts having completely different compositions and structures between species.

Layers

To understand the emergence of imprinting it helps to think in layers - a conservative minimum and a probable deeper root.

Conservative minimum (hard bracket): Because bona fide imprinting appears in both protostomes and deuterostomes, the capacity must be as old as their split, at least late Precambrian/early Cambrian, commonly estimated around 570-650 million years ago. That pins the origin of the rule (brief window + privileged channel + durable write) to the bilaterian ancestor.

Within vertebrates (elaboration layer): The fish-tetrapod bracket sets a second, younger minimum (≥ 420-450 Ma) for vertebrate-specific implementations - e.g., coupling olfactory templates to homing in salmon, or adding geomagnetic maps in turtles and some fishes. The amniote bracket (315-320 Ma) tells you the rule was already living inside the common ancestor of reptiles, birds, and mammals; the therian bracket (160-170 Ma) shows it continued unchanged in the line that produced every placental and marsupial.

Mammals didn't evolve imprinting separately, all inherited imprinting, resulting in a fundamentally shared trait. What changed are the channels and the parameter settings, not the existence of the genetic process as a trait.

IMPRINTED ORIENTATION

This book proposes that all sexual orientations stem from the same genetic process trait, and nothing in it should be misconstrued as delegitimizing towards, or critical of, LGBTQ+ orientations.

Evolution has no viewpoint. It is the sum of what persists. What was beneficial in an evolutionary context, is conceptually disassociated from what we as humans consider beneficial. Fingernails evolved to improve grip, now they hinder grip and are mostly used as decorations.

A woman doesn't consider the 'evolutionary intention' of fingernails when walking into a nail salon, because evolutionary intention doesn't exist. Evolution is not a guided process, and evolutionary intention is just a retroactive identifier of persistence origin wrapped into a convenience term.

Understanding our evolution, and the results of, is hard to do accurately. This is because we are a part of evolution, and thus, must view the forest from a tree. To understand evolution it's necessary that one adjusts for bias, as a scale needs to be calibrated for its own weight to accurately assess object weight.

An evolutionary unit for the purposes of measuring evolutionary impact might be a gene, an individual, a mating pair, a group such as a pack, or an entire species.

What decreases reproduction on a micro level might increase it on a macro level, and vice versa. Traits evolve in relation to each other, adding additional levels of abstraction to evolutionary benefit. We do not evolve merely on an individual level, our parts can evolve separately with us as an ecosystem and we evolve as a part of larger ecosystems. All evolution is self-referencing, that is to say, evolution only occurs within the context of evolution.

Lewontin's criteria clarifies what an evolutionary unit may be. There must be variation among units, consistent differences in reproductive success, and heritable transmission. Even entities without their own line of descent can function as vehicles for selection if they systematically bias the reproduction of the replicators they affect. A non reproductive person can still alter who reproduces and by how much, through impact on reproductive individuals. The purported group benefit of non-reproductive sexualities have formed the basis of existing theories on the origins of non-reproductive sexualities.

Theories such as the "Gay Uncle Theory," posit that

additional group-wide benefits delivered by a non reproducer offsets the reduction in individual reproduction. Theories of this nature tend to be possibility claims rather than probability claims, theories that exist because they are theoretically possible, not probable.

A theory which seeks to describe the genetic aspects of sexual orientation must explain the occurrence of non-reproductive sexual orientations in a manner which demonstrates how the evolutionary benefits of the genetic cause are substantial enough on a macro level to outweigh the decrease in individual reproduction, and thus persist in populations.

Imprinting as a process trait is the only theory which meets this standard. There has been much debate about the source of non-reproductive sexual orientations. This debate is usually framed as nature versus nurture, or genetic versus social factors. The broader landscape of fixed, idiosyncratic sexual orientations reinforces a process-based explanation.

Genetic Evidence Lacking

Although sexual orientation shows non-zero heritability, twin studies consistently report incomplete (<30%) non-heterosexual concordance among monozygotic[5] twins.[6]

[5] Monozygotic twins are those with 100% identical genes.

[6] This figure intentionally excludes pro-banded and other concordance-inflating studies.

This fact is striking. Genetically identical twins, usually reared in the same household, are unlikely to share the same sexual orientation. This appears to defeat both sides of the “nature vs nurture” argument.

30% non-heterosexual concordance means that if one twin has any non-heterosexual orientation, there is about a 30% chance of the other twin also having a non-heterosexual orientation. This 30% doesn’t indicate *same* orientation overlap, it merely indicates *non-heterosexual* overlap. Non-heterosexual is a broad category.

It doesn’t require much education to understand how unlikely it is for a non-reproductive sexuality to be a direct result of a gene or set of genes. It’s harder for a gene that encodes non-reproductive sexuality to be passed on via reproduction.

This elephant has been in the room for some time, and shoehorned attempts at justifying genetic non-reproductive orientations such as "The Gay Uncle Theory" fall quite short.

Besides, there are many more sexual orientations besides being merely "straight" or "gay,” each observable to be a real biological impulse, not mere preference. While there are those who believe that being straight or gay is a genetic trait, few would argue that an uncontrollable sexual compulsion to, say, have sex with a toaster, could be genetic in origin.

Yet to a person with an unwanted object- oriented sexual compulsion, his sexuality is just as "real" or "set" as any other sexuality.

Take Edwin Tobergta, who in 2013 received his fourth conviction for having sex with an inflatable object. Whether an inflatable pool toy floating seductively in a public pool, or his neighbor's inflatable pumpkin, Mr. Tobergtas object-oriented sexuality has landed him in jail four separate times over the decades. This man isn't "faking it," his sexual orientation is as real as anyone else. As inflatable ornaments did not exist historically, it's unlikely his sexual orientation stems from direct expression of evolved genes.

There are thousands of niche sexual orientations, many with their own active community. People can dedicate their entire lives to acting out their niche sexual orientation, and many have died or gone to prison after being unable to suppress their niche sexual desires.

These orientations are real, yet the majority of these sexual orientations couldn't be genetically encoded, being related to objects that didn't exist on an evolutionary scale. The large communities that exist for niche object-specific sexual orientations leaves random genetic mutation an unlikely cause.

This conundrum is solved by understanding sexual orientation to be the result of a genetic trait, that of imprinting. The portion of reproduction lost to non-reproductive sexual imprints is offset by the increased chance of remaining in an evolutionary lineage. Individual lineage terminations are justified by the enormously increased ability of lineages to accept beneficial mutations. Imprinting facilitates evolution itself, an evolutionary advantage bordering on a requirement, and imprinting as a process trait persists despite its imperfect expression.

GENERATED INPUT

Every study on the subject has shown that when input-processing circuits of the brain lack external signals, the brain generates fake input. Understanding the tendency of the brain to generate input when actual input ceases, is essential to understanding how imprinted engrams are later referenced by the brain.

Every Study Agrees

In the mid twentieth century at McGill University, Donald Hebb's team restricted volunteer visual, auditory, and tactile input with goggles, gloves, and continuous white noise. Within a day or two, many reported vivid imagery, perceptual distortions, and disorientation. Most could not continue.

Around the same time at the National Institute of Mental Health, John C. Lilly built the isolation tank to study how the mind behaves in near total silence and darkness. In his report, "Mental effects of reduction of ordinary levels of physical stimuli on intact, healthy persons," he describes the vivid imagery and altered states that emerge when the usual sensory stream is cut off.

More recently, Mason and Brady's 2009 anechoic chamber study showed that even fifteen minutes of sensory deprivation can result in hallucinogenic experiences, including perceptual distortions and paranoia, in healthy participants.

Follow-up work by Daniel and Mason replicated and extended these findings while controlling for anxiety, suggestibility, and fantasy proneness. Deprivation still resulted in psychotic like experiences in both high and low hallucination prone groups.

The mind supplies its own signals to maintain neural architecture when external stimulation is absent. Without input, circuits are reassigned or begin to deteriorate. The brain does not power down like a computer. We have all experienced this generated input during sleep, when actual sensory input ceases, the brain generates input that we call dreams.

Regardless of the evolutionary purpose this generated input serves, clearly, there is generated input.

This internally generated input most likely exists to help maintain neural structure in input processing regions of the nervous system, when there is nothing to process. Input is required to maintain the neural structures that process input. In a manner analogous to how water shapes a river, even as it is held by a river, the input our brain is designed to receive shapes neural pathways even as that input is constrained to the same neural pathways.

People with eye disease or damage to the visual pathways can perceive vivid, silent images such as faces, patterns, and scenes despite the absence of external input. This reflects the visual system generating content.

Sudden hearing loss coincides with generated audio input. Beyond simple tinnitus, some people with hearing loss perceive music or voices that are not present. These are internally generated percepts that emerge as auditory input diminishes. Reviews report meaningful prevalence among hearing impaired groups.

After amputation or nerve injury, many people continue to feel the missing limb, often with pain. This is not 'generated input,' usually neural misallocation during neural reallocation, but there are blurred lines. Neuroimaging links these sensations to active neural misfires in addition to neural reorganization.

Following infections, trauma, or degenerative disease that reduce real olfactory input, some individuals similarly perceive smells or tastes that are not present.

Let's take a look at what happens in the brain when input is ceased, to understand just how rapidly reallocation happens, and how desperately important the generated input is to prevent neural reallocation.

Visual Input Cut

<2 hours.

Occipital networks start shifting their coupling and responsiveness during touch tasks after just 2 hours of blindfolding. Most changes centered around dorsal visual areas. Non-zero neural reallocation happens within just two hours.

Days 1-2

With continuous blindfolding, visual cortex excitability rises and most participants report complex visual hallucinations by day 2.

Day 5

After 5 days of total blindfolding, primary visual cortex shows increased response to touch and Braille performance improves; brief rTMS[7] to occipital cortex disrupts Braille in the blindfolded group only, demonstrating neural reallocation from visual processing centers of the brain to other tasks. The changes are rapidly reversible at this point.

Weeks - months

Visual cortex repurposed to process other input. V1 and extrastriate areas become reliably task-positive for touch and sound, such as Braille reading, tactile shape, auditory localization, speech/language parsing, even verbal memory.[8]

---

[7] Repetitive trans-cranial magnetic stimulation - interrupts a brain region temporarily.

[8] Disrupting the visual cortex now affects other brain functions, demonstrating that the visual cortex has been repurposed.

Years
Occipital cortex shows much improved function in those who went blind as children, rather than in adulthood.

Crucially, there is also around 25% less gray matter remaining after neural reallocation of the primary visual cortex to processing other input. This shows a "reallocation cost" that is paid for neural plasticity, explaining why reallocation is desperately prevented instead of embraced.

Degeneration to the occipital cortex is much more pronounced in those blinded in adulthood as opposed to childhood, demonstrating reallocation and not misallocation causes the neural damage. It is the change, not the location, that impacts processing ability.

The brain can reallocate neural resources, but as each reallocation results in reduced function, neural reallocation is the last resort of the brain. Generated input can maintain input processing pathways long enough to prevent reallocation.

Audio Input Cut

<7 days
Brainstem and cortex attempt to up-regulate "volume" through reduced inhibition, higher spontaneous firing. Risk of tinnitus, perceptible brain generated audio input, rises. Auditory cortex shows less stimulus-driven activity, more intrinsic activity.

7 - 30 Days

Start coupling more with visual and somatosensory networks, coupling with thalamic auditory relay (MGB) wanes.

6 Months
Pathway changes. Emerging degeneration in the peripheral/auditory pathway, such as the auditory nerve, brainstem inputs. Better lip-reading, peripheral visual attention, and tactile temporal acuity.

> 1 Year
Structural remodeling with gray-matter loss in primary and secondary auditory cortex, reduced integrity of auditory white-matter tracts; thalamic volume declines. Auditory cortex becomes task-positive for motion, face recognition and touch, resting-state networks rewire around this new role. As time goes on, peripheral auditory structures continue to atrophy.

Impact on the brain from the loss of taste is less studied, those losses being very rare and less impacting, yet the studies that do exist show similar neural reallocation and associated reallocation-related degradation.

The reallocation of sensory-specific brain circuitry upon cessation of sensory input is less dramatic than the reallocation of brain circuitry unpaired with such 'hardware.' The portions of the brain that process higher order input, such as language and social input, demonstrate much more plasticity than, say, the occipital cortex.

As a rule of thumb, older-evolved segments of the brain

show less plasticity.

Brainstem, hypothalamus, thalamic relays, and primary sensory cortex display diminishing and more tightly bounded plasticity in ontogeny, whereas association areas and prefrontal cortex retain broader plasticity that persists into adulthood.

It is clear that the brain generates sensory input when real input ceases. Yet there are more input processing regions of the brain, than exists to process input from our three distal senses. Given the greater plasticity of areas associated with social input such as language, as opposed to neural regions associated with a particular sense, it is reasonable to infer that generated input might be required to maintain those structures in absence of the expected social input.

IMPRINTING IN HUMANS

Human imprinting processes should be studied, mapped and meticulously described on a genetic and observational level. Since they're not, the observation-originated models of human imprints in the following chapters are based upon conjecture, provisional theories and unsubstantiated logical inferences. These models are incomplete at best, being new, but they are presented to the educated reader for consideration and acceptance of the portion believed.

As my old magister used to say, there are two types of people, those who can learn from incomplete datasets.

Homo sapiens have a long developmental period compared to other mammals, and imprint phases and stimulus across this period are harder to identify than in other animals.

Three imprints in Homo sapiens are proposed, with the best intentions towards, but no assertion of, accuracy.

This is a <u>theory</u> book, the following are proposed models.

Mother Imprint

Sensitive Period: Infancy, during weaning.

Evolutionary Object: Mother.

Identifying aspect: Breastfeeding, source of.

Engram: A lactating woman is imprinted on, creating a 'mother engram' which creates early recognition of the mother. This engram is stored and later referenced by the nervous system as the form of a reproductively successful woman in an animal's space and time.

Alpha Imprint

Sensitive Period: Post-fear periods as they occur between age four to 25, tracking α2A receptor activity required for 'depth' of imprint, which peaks around age 14.

Sufficient fear temporarily makes the imprint substrate accessible. With sufficient norepinephrine, the end of the inverted U that represents the correlation between fear and α2A activity is reached, severing dorsolateral PFC firing, shifting input along dorsal striatum → globus pallidus → thalamus → to sensory association cortices where Hebbian writing records the engram.

β-adrenergic/CRF signaling in amygdala and bursts from SNc in dorsal striatum capture PRPs and associate threat value and stimulus together into an engram.

Evolutionary Object: Troop alpha.

Identifying aspect: Fear, source of.

Engram: The personality of the alpha is recorded, with each successive engram stemming from an occasion of fear, representing a group parameter. Collectively these engrams accumulate to form an alpha personality, an engram representative of the personality of the alpha.

Sexual Imprinting

Sexual orientation[9] is derived by the nervous system in reference to the mother template, positively in male and inversely in females.

---

[9] Only sexual orientation, not gender perception or similar.

MOTHER IMPRINT

Across species, the mother is the default target of early imprinting. This imprint helps the developing brain recognize its mother. The resultant 'mother engram' represents both the mother's form and the template for a reproductively ideal woman within an animal's particular space and time.

In mammals, early recognition of the mother has been demonstrated to permanently impact perception, as shown in studies on sheep-goat cross-fostering. Newborn lambs reared by goats, and kids reared by sheep, later show preference towards the foster species for the purposes of mate choice and social recognition.

In ewes, the effect is laid down at birth through maternal contact and odor. Studies have shown that if β-adrenergic receptors are blocked during this imprint window[10], imprinted recognition fails in ewes. Similarly, if you lesion centrifugal noradrenergic input to the bulb, ewes do not form a selective bond and may accept alien lambs.

This interesting study lends itself to the consideration that certain medications might have an impact on the imprinting processes in primates.[11]

Other mammals have olfactory imprints. Although Homo sapiens don't have any clearly defined olfactory imprints, it's likely smell plays a role in early pre-conceptual object recognition. In Homo sapiens, infants are born functionally blind[12], and auditory cortex continues maturing for years, by contrast, olfaction arrives ready to drive survival behaviors from day one.

Unlike input from the other four senses, which must first be sorted through the thalamus, neonatal smell runs directly from the olfactory bulb to the amygdala and

---

[10] e.g., via propranolol perfusion.

[11] Many common medications are adrenoceptor-affecting, such as phenylephrine which is the most common addition to acetaminophen (Tylenol). Currently, studies associate Tylenol intake in gestating women with increased risk of autism, although nobody knows why. **If** negative impact on social imprints is the cause of irreversible social dysfunction that can't be explained genetically, such as in autism, understanding imprinting could help prevention of such.

[12] Acuity often quoted below 20/600

hypothalamus to directly impact early reflexes and emotional association.

Irrespective of recognition pathways, it's likely that Homo sapiens have the same imprint as exists in most animals whose imprints have been studied, that of the mother.

ALPHA IMPRINT

Humans are "pack animals." The U.S. Surgeon General's 2023 advisory puts the overall mortality impact of limited social interaction on par with smoking 15 cigarettes a day. The increase in mental disorders such as depression and dementia is even more pronounced than the incredible increase in mortality.

Yes - according to the world's foremost health experts, you are better off smoking a pack a day, or drinking 42+ beers per week, than being lonely.[13]

If you (1) live alone, and (2) have no regular group participation (church/club/sports) and (3) have very infrequent contact (less than monthly) with

---

[13] U.S. Department of Health and Human Services, Office of the Surgeon General. 2023. Our Epidemic of Loneliness and Isolation. https://www.hhs.gov/sites/default/files/surgeon-general-social-connection-advisory.pdf

friends/relatives, you are in the "isolated" range, and might die of loneliness.

That's for relatively ordinary social isolation. When a human being is cut off from other primates completely, the mind, which evolved to function as one component of a group of minds, starts breaking down instantly.

Within a couple of days into isolation, the brain slowly begins generating false input, and confusion alongside permanent brain damage begins to set in. Merely being separated from fellow humans, into an otherwise safe room, causes rapid accumulative neural malfunction. Those who have spent continuous time in a 24/24 jail cell – including myself for 8 days – can attest to the rapid onset of hallucinations.

Homo sapiens are the exemplar social animal, and our nervous systems have evolved to guide us as part of a group. A glance at the evolutionary differences between humans and other primates shows a clear trend towards integration with the group.

Compare a human's eye to that of a gorilla. A gorilla's eye is pitch black - from sclera to pupil. A gorilla has no "white of the eyes." This evolution conceals a gorilla's gaze from others, making it harder for other troop members to ascertain directional intention, an advantage for the individual, against its peers. It is very hard to see where a gorilla is glancing, and staring directly into a gorilla's eyes will spark a rage reaction.

Gorilla's eyes evolved to hide their intentions from each

other, whereas Homo sapiens have evolved eyes to signal intention to each other.

Human eyes constantly broadcast both directional attention and emotional state to other members of the troop. Even when you can barely see a person in the distance, you can ascertain that person's directional attention from their eyes. Hiding one's thoughts and emotions from others is a very rare skill that takes great practice. We are built to broadcast our emotions to others, towards group benefit and against individual benefit.

Homo sapiens are *the* group-defined primate. Homo sapiens share child rearing, divide labor and pool resources, and coordinate via the most advanced communication of any animal.

Most of the evolutionary variances a human has from other primates are towards greater integration with the troop. Given this, it stands to reason that processes which instill group structure would be more pronounced in humans, not less.

All primate groups have a group structure. A group structure does not imply a hierarchical group structure, although hierarchical group structures are most common. Most primate groups are under control of an alpha. The common misconception that most primate groups are *not* under control of an alpha, stems from a misunderstanding of what an alpha is.

Dominant females in lemurs, bonobos, callitrichids, and

baboons are still alphas. An alpha still controls a troop even if members are semi-solitary.

In Homo sapiens closest relative, chimpanzees, alpha status does not require physical dominance at all. Jane Goodall once described a scenario where a physically weak chimpanzee named Mike became an alpha, using only sound to terrify.[14] For chimpanzees, there are many paths to alpha status besides direct physical violence, as chimpanzees understand deferral of power.

A wild chimpanzee named Wilkie was observed to have 'won the popular vote,' grooming his way to alpha status over larger opponents. Diplomatic paths to alpha status are common, with smaller chimpanzees achieving alpha status through violent coalitions against larger troop members, or via collection of sufficient deferred power from troop members.

In orangutans, early life determines whether a male becomes a "flanged" (cheek pads + throat sac) or "unflanged" orangutan. There's no genetic difference between flanged or unflanged orangutans, yet the males will develop completely different physical features based on group status alone.

When a marmoset or tamarin female is subordinate under a dominant breeding female, social cues suppress her ovulation until her status changes.

---

[14] An interesting read: https://www.nationalgeographic.com/magazine/article/gombe-chimpanzees?

When a Mandrill male becomes the troop alpha, his red-blue face and genital colors intensify and his testes enlarge, but both return to normal if he drops in rank.

When a Gelada male becomes a harem leader, his breast "bleeding-heart" patch flushes brighter red and larger in step with his group status.

The vast majority of status related changes in primates are behavioral, and not physical. Physical changes in response to social rank are mentioned here only to highlight how important group status is to primates.

The physical changes based on group status in other primates is very telling of a key concept. Recognition of the alpha of a primate troop is evolutionary important enough to the troop, for physical identifiers of rank to evolve. Primate troops, generally speaking, are structured so that each member serves a different function based on status.

If members of a primate troop are not bound to the groups parameters, the troop ceases to exist. Without an adhered to structure, the troop devolves into separate primates with individual interests. The alpha of a primate troop does not just help the troop, it creates the troop.

The expressed personality of the troop alpha establishes behavior parameters, and in moments of imprint vulnerability this personality is imprinted upon troop members. The engram of the alpha is a persistent template of group parameters.

Just as a mother imprint orients a primate toward its mother and future mate, an imprint is also needed to orient Homo sapiens to their troop.

These primate group orientations and statuses are innate, and cannot be easily understood through logic. The highest IQ person can struggle to conceptually understand even the clearly defined social customs of modern society. In fact, studies have shown the more intelligent someone is, the more they struggle to socialize.

Cumulative alpha imprints, each representing a group parameter, together create a template of the troop alpha, which serves as a standard to be referenced for group parameters. This alpha template is referenced for dual purposes, with expression dependent on the eventual group status of the individual.

The interests and parameters of the imprinted alpha are expressed as personality, similar to how our own sense of self is displayed internally as personality. We know that the brain references a series of engrams[15] to create our individual personality.

The same systems can reference the alpha engram, to generate social input from the alpha personality, when normal social input is ceased. This helps align individual motivations to that of the group.

Upon cessation of actual social input from the troop, the

---

[15] In the case of individual personality, these engrams might be memories or similar.

brain generates social input from this alpha engram. This ensures key parameters of the troop alpha are enforced, even when the troop isn't physically present.

Before the form of an alpha is imprinted for later input generation, the object must be identified based on a signal indicating type.

What indicator has evolution decided to be the mark of the troop alpha, as certainly as breastfeeding is the mark of the reproductive female?

Unfortunately, fear.

In primates, alphas use force to create group parameters which are later respected through fear. Even in modern Homo sapiens, what defines a government is a monopoly on the use of force.

When a developing person experiences real fear[16] towards others, it initiates an imprint phase until cessation of fear.

This imprint phase is repeatable, not alterable. Each fear phase results in an engram stored separately from normal memory, to be referenced when constructing the alpha personality. Each fearful occasion represents a group parameter. The cumulative imprinted parameters of the alpha(s) represents the parameters of the group, and these occasions are referenced for the purposes of social input generation.

Group status determines expression of this alpha

---

[16] Not similar emotions such as disgust or disdain.

personality.

High Status Members

In higher status Homo sapiens, the aggregate alpha engram is referenced when constructing individual personality, with the result of merging the previous group parameters of the old alpha male, into the new. The templated personality is incorporated and expressed outwardly.

In socially isolated individuals, the alpha personality becomes a source of generated social input experienced inwardly as external perspective, and not as a part of individual consciousness. This low status expression results in an individual proportionally beholden[17] to the perspective of the alpha personality. Virtually every person cut off from society, especially the homeless, will start arguing with this imaginary personality.

The high status, outward-facing expression of alpha personality traits becomes more pronounced as a person becomes higher status. Similarly, the more isolated a person becomes, and the less actual social input they receive, the more input is generated from the templated personality.

While 'group status' seems an unscientific descriptor, it's the correct one. The primate brain knows its status in the troop. There's no argument against status recognition in Homo sapiens that does not apply to other primates.

---

[17] Proportional to their degree of separation from troop.

The brain increasingly relies upon input from the alpha template in times of reduced or ceased social input, conceptually similar to how the nervous system generates sensory input when a sense is cut off.

The nervous system represents the desires of the alpha in a similar manner as it represents the desire of the individual, through consciousness.

The 'reason' the brain represents the desires of the alpha as consciousness, rather than relying on conceptual interpretation of remembered group input, is for the same reason we use the same eyes to see ourselves and other people, the same hands to hold our stuff to hold other people's stuff, it's just what we have for the specific function.

Evolution is lean and doesn't often lend different hands for different tasks. The brain references at least a set of engrams to represent one's own cumulative experiences as consciousness, and the same system references separately stored cumulative experiences to represent alpha perspective. Consciousness is the chosen format of the brain to represent personality, there is no other I am aware of.

Generated social input begins with isolation and ends with sustained reintegration.

The alpha circuit affects social interaction. Every group of mammals sorts itself into pecking orders with mostly nonverbal signals. You see it in a puppy litter within days, you see it in a conference room within minutes.

Those individuals who grew up missing the type of fear-based social input which nervous systems have evolved to expect, as occasionally occur in wealthier households with very limited exposure to conflict, can experience social dysfunction.

These individuals are "booksmart," but can't intuitively comprehend even the most basic primate interaction. These individuals have to awkwardly wield logic as a tool to understand otherwise innate human behavior. They can't navigate other humans without assistance. These are those individuals who wouldn't 'survive in the wild,' for lack of a better descriptor.

Cognitive function in Homo sapiens evolved from the addition to, and repurposing of, older nervous system components which evolved to serve purposes we no longer have. There are many impulses and conceptualizations which are inherent to primates and can't be usefully understood by conceptual logic. The remnants of our primate past are reflected in language.

As an example, in all 7,100 languages spoken on earth without exception, "up" means good, "down" means bad. This is an innate association, not a logical association. It's actually easier to move downwards - climbing up is hard.

Moving *up* in a company means moving to a *higher* position, which means a better position. If you're feeling *down*, you are depressed. If you feel good about yourself, things are *looking up*. If you are elated from drugs, you're *high*.

This association is from our evolutionary past, as primates evolved on trees, and higher status primates took the geographically higher position. Therefore we experience social motivation as upwards. An autistic person has to mentally map, using conceptual logic, these otherwise innate spatial-associations. They don't intuitively understand why, say, giving unsolicited advice to someone would cause insult. The autistic person has no inherent conceptualization that his advice implies the person is below him, as directives in primate troops are normally given "downwards."

Similarly, an autistic person struggles to understand the importance of swear words, because they don't 'feel' the associative impact. Words evocative of feces or genitalia don't evoke an associative emotional reaction in an autistic person, because the implication of these words are derived from the objects inherent meaning in a primate troop context.

In a troop where feces are used to mark territory, and sex acts are crucially important, a word that invokes feces or genitalia triggers an emotional reaction, in the absence of the physical object, because the item has inherent primate rules associated with it, which are violated by out of context invocations. The unauthorized wielding of the symbolic item, out of its innate primate context, is experienced by the socialized primate as 'wrong', like being called by the wrong name. It causes discomfort.

We experience remnants of our primate past every day. If you are one of the many people who have experienced a sudden falling sensation while nearly asleep, and spread your arms out to prevent an imaginary fall, then you have experienced an ancient reflex originally intended to prevent your fall from a tree branch.

Our nervous systems are still partially serving our past iterations, and the mechanisms of such are too specific for someone to function in society with only a logic-derived understanding of other people. The nervous system evolved to expect certain kinds of input, and without that input there is generally malfunction.

Like-Kind Expression

When behavioral urges affected by the alpha imprint outwardly express in high status socialized Homo sapiens, this expression is usually like-kind. The personality is partially integrated, the expression is various.

In someone whose most significant childhood fear is public humiliation, there may develop in adulthood a compulsive drive to humiliate others. This is the same personality trait being expressed, although exactly how it is expressed might vary enormously.

That is not to say that exact expression doesn't occur. For example, a child molestation victim is 3.3x more likely to molest a child. Convicted child molesters are 7.1x more likely. That means being a victim of child molestation is 46% as strong an indicator of being a child molester, as having previously molesting a child. These are not mild associations.

Even if someone was to disagree completely that 'fear results in imprints which are then incorporated into personality,' nobody can deny that traumatized people are much more likely to traumatize others in similar ways. The association between being hurt, and hurting others in a similar manner, is not disputed by anyone, and shows up in every study.

This like-kind expression emulates the real or imagined intentions of prior alpha(s).

Religious Impact on Developmental Imprints

Religion has always diverted human emotion and development from tangible, perceptible entities towards entities unseen.[18] This impacts behavioral perception.

It's possible that religious upbringing might impact behavioral perception in ways previously unanticipated, through the alteration of imprints.

In many religious upbringings, emotions and cognitive function in general can be affixed more primarily upon entities unseen than perceptible troop members, ceding the developmental impact of troop members to these representations.

A developing Homo sapiens, experiencing much greater fear towards these representations of troop members than towards actual troop members, might imprint upon the representations instead of ordinary troop members,

---

[18] This is not a critique of religion.

especially in relatively isolated households.

This could partially explain why religious groups have outlasted other secular associations. Standardization of the alpha imprint could provide a better imprint than would otherwise be received from other primates.

For example, in the monotheistic religions which have dominated the world, the main entity that children are directed to fear is either Satan or God.

Let's consider Satan as an imprinted alpha, and imagine significant fear of Satan results in an engram of his personality as depicted.

So, how is the personality of Satan depicted?

His very brief appearance in the Torah/Bible[19] is to tempt humans to eat from the Tree of Knowledge. Satan's knowledge-seeking personality is one many parents would like their child to have. The negative impact of his action, that of tricking humans into biting from the Tree of Knowledge, is only contextually derived. No negative effect stems from his personality directly, all violence is from others. Satan himself is depicted as entirely non-violent, persuasive and singularly knowledge-seeking in his appearance. As a representation of alpha personality, Satan appears a much better choice than that of your average alpha.

---

[19] There are brief mentions of Satan besides Genesis, but these are not taught to children, and thus irrelevant.

If the imprint created from fear of a religious personification instead of an actual primate results in a more beneficial imprint, it stands to reason that those imprinted as such would have advantage over other people, and groups whose members have this imprint would be more likely to outperform and "reproduce" their ideology.

Ideologies that provide a benefit to imprinting can have a survival advantage, being adapted to an evolutionary process. They are more likely to persist over time.

To a member of a monotheist religion who is raised to fear Satan in this specific manner, there might be personality change to group benefit. In increased status, a knowledge-seeking personality is expressed outwardly, encouraging others to knowledge. In lowered status, a knowledge seeking imperative is experienced inwardly from social input generated from this engram.

The focusing of developmental fear on Satan is most strongly retained in Judaism[20], and significantly retained in Catholicism and many other branches of Christianity.

This effect could partially explain the disproportionate contribution of Jewish peoples towards scientific and creative industries, relative to other groups.

Other religions have seemingly created standardized imprints throughout history, by directing emotions and

---

[20] I was enrolled in both Jewish and Catholic early education, and subjectively experienced this focus.

action towards representations of persons instead of local persons.

Of course, for many raised in monotheistic religions, the primary entity feared during development is God, which would result in a different imprint under this interpretation. Individuals who primarily feared Satan would perhaps trend towards knowledge-seeking, concerned with pursuing some 'truth.' Those more fearful of God might trend towards standard enforcing in relation to what 'is', more focused on who 'deserves' what. Or, perhaps some blend of both.

Negative Expression

The most intentional upbringing can be interrupted by unexpected trauma. The internalization and projection of this trauma is apparent even outside of a developmental context. Trauma occurs when the negative aspects of a bad experience overwhelms the minds ability to process it.

The brain does process negative experience, just not always in the way we would like. From an evolutionary perspective, the most negative experiences in our lives can be the most relevant for the nervous system to bring to the forefront of the mind.

If a lion attacks the troop during foraging, it makes sense that this lion attack be constantly brought to the surface whenever the troop is foraging. It is not as helpful for the memory of, say, a violent rape to be resurfaced, whenever a person enters a setting similar to that of the original attack.

With conceptual logic sufficient to determine the majority of avoidance-decisions in our relatively standardized society, the ancient emotional-association avoidance systems can serve merely to amplify and continue trauma, sans benefit.

Trauma experienced as a child persists into adulthood, as these fearful experiences are recorded separately from normal experiences and referenced for the purposes of social input generation. In low status expression, an individual might experience compulsive behaviors that adjusts their actions towards the benefit of their previous abusers, leading them to both seek out and make themselves more available to similar abusers. This is a result of the abusers actions being recorded separately during fear states as an alpha engram, which are later referenced by the nervous system to generate the personality of the alpha.

The impact of alpha personality 'perspective' manifests as a constant stream of influenced decisions, adjusted towards the motivations of the alpha personality, irrespective of conflict with personal motivations.

The fate of those with bad imprints on this 'alpha circuit' can be quite cruel, if they do not belong to a group. A member of any group, who has status and regular social input, never has social input generated from the alpha personality, and is thus free from 'competing motivations.'

Without any actual perspective from others, group

perspective is represented by alpha personality, ensuring continued adherence to the group even in the absence of others.

That's part of why the most important thing for any abuse survivor is socialization, and therapy. Healing in therapy starts just by having someone who cares enough to provide input into your life decisions, even if no useful suggestion is made. Just talking begins the healing process by replacing generated input with real input.

Religion, the primary source of action towards that which is understood but undefinable, has long facilitated this absolute need for social input, and religious confession provides an opportunity for members to receive input and advice on even the most secret experiences which would otherwise alienate.

SEXUAL IMPRINT

Sexual orientation in Homo sapiens is derived from the engram of the mother which is imprinted upon a child during nursing. This engram stores her attributes as that of the ideal reproductive female in his space and time.

Females receive the same imprint of the mother, however their nervous system references this mother template inversely, away from and not towards attributes of the mother in relation to troop averages.

Non-reproductive orientations occur when someone or something besides the mother is imprinted as the mother.

These non-reproductive imprints result in the extraordinary range of human sexuality.

ABOUT THE AUTHOR

Max Freedom Pollard is a pioneering author, known for his ethological and etymological work.

amazon.com/author/maxpollard

wikipedia.org/wiki/max_freedom_pollard

instagram.com/max_freedom

linkedin.com/in/maxmpollard

EVIDENCE PENDING
An Epilogue About Bad Data

Studies on environmental factors affecting later sexuality are very controversial, and it's a category of study deeply impacted by political and social belief. None of the existing data has been interpreted so as to be supportive of the theory of sexual imprinting, or any other theory of sexual origin for that matter.

A barrier to study is the belief that the origin of sexuality is related to how 'moral' a sexuality is.[21] Studies with the explicit purpose of attempting to determine the origins sexual orientation are all but banned in academia unless subsumed by some different intention.

The most advanced gene sequencing and analysis have shown no evidence of a gay gene, yet inversely, studies on environmental factors show next to no impact on orientation.

---

[21] More commonly, the belief that other people believe this.

Current data is generally supportive of no existing theory at all. As this book posits a novel, but relatively static, origination of sexual orientation - that of the engram of a breastfeeding mother - an examination of existing studies on breastfeeding and sexual orientation is required.

As of 2025, there exists merely two relevant studies comparing the impact of traditional breastfeeding on the development of traditional sexual orientations. There is the 2019 study on the topic, and a 'follow up' 2020 study conducted by different researchers to be published in the same journal.

Analyses of these two studies demonstrate some of the problems inherent to deriving *cause* of result, and necessarily cause of associated factors, from comparison of associated factors. This is a common problem in etiology. This is less a problem with raw data, enormously a problem with heavily adjusted and curated data such as in these studies.

A clear linear association is found in the first study, albeit with degree reduced as close to zero as possible through adjustments, while none is found in the 'follow up' study, which appears designed specifically to negate the first.

Measuring The Impact of Breastfeeding

If sexuality is derived in humans from a "mother imprint" during breastfeeding, one would initially expect to find a significant association between bottle feeding and non-reproductive sexuality.

The problem with interpreting both studies for our purposes, is that adjusting for other factors related to non-reproductive sexuality only improves accuracy when comparing breastfeeding to other factors, having the opposite effect when breastfeeding is examined as a cause of factors.

Every adjustment for factors already associated with non-reproductive orientations, reduces the association between breastfeeding and non-reproductive orientations by the amount adjusted. This destroys 'degree,' the extent of impact.

The impact of any cause, is necessarily reduced in relation to the number and impact of all factors downstream of the cause.

Crucially, I note that having a non-reproductive sexuality would have middling correlation with having a non-reproductive sexuality, if having a non-reproductive sexuality was a factor in these studies.

Although the degree of impact has been shrunk, the trend remains. A reduction in non-reproductive sexuality in those breastfed, with the reduction increasing linearly with each additional month breastfed. A clear correlation without deviation.

The two main studies on environmental factors that result in non-genetic sexuality are fundamentally flawed, and not just for our purposes. They're also just bad studies with bad methodology, on their own merits. The first is palatable, the second gratuitously corrupted. I analyze them here, only out of an obligation to address existing data.

ALSPAC cohort (UK), 5,007 teens — Developmental Psychology (2019)

This study measured breastfeeding duration in infancy and sexual attraction at age 15.5. For boys only, shorter breastfeeding duration was associated with slightly higher odds of later non-heterosexual orientation after adjusting for many factors. Effect size was small (odds ratio about 0.90-0.92 per additional month of breastfeeding), and absolute risk differences were 1-2 percentage points.

In boys (n=2,349): Each additional month of breastfeeding was associated with lower odds of reporting a non-heterosexual orientation at 15.5y; OR = 0.909 (95% CI 0.838-0.985).

This study confirms an increased rate of non-reproductive sexuality in bottle-fed boys, with each additional month of breastfeeding resulting in a reduced chance of a non-reproductive sexuality. No difference is shown for girls.

The linear reduction in non-reproductive orientations in breastfed infants, with each additional month increasing the reduction, establishes the impact of breastfeeding on sexual orientation in males, even with degree artificially removed.

Millennium Cohort Study (MCS), 9,795 teens (14y) — Developmental Psychobiology (2020)

MCS is a follow up to the first ALSPAC study. Despite being a follow up study, MCS has every inadequacy of the first study with far worse methodology.

MCS measured attraction at age **14** using two separate items, as opposed to ALSPAC's 5-point attraction scale at age 15.5. Even accepting the presumption that non-heterosexual orientations are usefully self-reported by 14 year olds, the MCS also added a "mischievous responding" screening co-variate to soak up adolescent responses they deem "implausible," causing further damaging to already inadequate data.

MCS adjusted for maternal psychological distress, parent-child relationship, family socioeconomic position, house moves, and pubertal BMI in addition to many other factors both tangible and esoteric.

Several of those factors are tightly correlated with breastfeeding itself and their control further reduces degree.

In ALSPAC, mean breastfeeding duration for boys was 6.1 months, giving a reasonable gradient. In MCS it was much shorter on average (3.6 months for boys), increasing noise and reducing linear contrast.

Both studies had incomplete datasets and methodology, and used chained-equations multiple imputation, with ALSPAC up to 57% missing on some variables and MCS up to 50%.

These studies are really bad. Perhaps they are bad by

design. Many universities shy away from studying non-genetic causes of sexual orientation, wary that such research could be perceived as offensive to LGBTQ people.

It is hard to overstate just how reluctant those in most academic environments are to conduct research on these topics.

As far as I can ascertain, this reluctance, based on an imagined reaction from the LGBTQ community, is born out of homophobia. I have never encountered an LGBTQ individual who I would presume to take offense at such a thing.

Evolution, and products of, are entirely unrelated to morality and validity. 'Evolutionary purpose' does not exist outside of retrospective biology. My hands might have evolved to grip tree branches, but I use them to write unsubstantiated ethological theories. In my entire life, I have never met a person who conflated evolutionary purpose with morality or validity. I hope there will be better data on this topic as academia matures.

www.ingramcontent.com/pod-product-compliance
Lightning Source LLC
LaVergne TN
LVHW090426210926
848480LV00002B/706